BROWN BEARS
AT THE ZOO

written by Mia Coulton
photographed by Mia Coulton & Amy Musser

There are many **bears** to see at the zoo.

This is a **brown bear**.

Brown bears are very big. They can weigh over 700 pounds.

Brown bears are very fast. They can run up to 35 miles an hour.

This bear has fur with grayish tips.
It is called a **grizzly bear**.

Grizzly bears are brown bears.
Not all brown bears are grizzly bears.

Brown bears are **omnivores**.

They eat both plants and meat.

The front **claws** of a brown bear are very long and sharp.

They use their claws for digging and to catch food.

Brown bears have a good sense of **balance**.
This grizzly bear is walking on a log.

Brown bears can climb trees when they are young.
As they get older and heavier, it becomes harder for brown bears to climb trees.

This brown bear is having fun in the water. Brown bears are very good swimmers.